THE LIBRARY OF WHY?™

Why Is the Sun So Hot?

Patricia J. Murphy

The Rosen Publishing Group's
PowerKids Press™
New York

To the sunshines of my life: E.A., O.B., J.O.M., D.J.M.,
A.M.A., E.B., M.A. and B.B.

Published in 2004 by The Rosen Publishing Group, Inc.
29 East 21st Street, New York, NY 10010

First Edition

Editor: Frances E. Ruffin
Book Design: Danielle Primiceri
Layout: Nick Sciacca

Photo Credits: Cover, pp. 4, 8, 19 © Digital Vision; pp. 7, 15, 20 © Photri Microstock, Inc.; pp. 11, 12 © PhotoDisc; p. 16 © Lani Howe/Photri Microstock, Inc.

Murphy, Patricia J., 1963–
Why is the sun so hot? / by Patricia J. Murphy.— 1st ed.
   p. cm. — (The library of why)
Includes bibliographical references and index.
ISBN 0-8239-6235-0 (library binding)
1. Sun—Miscellanea—Juvenile literature. [1. Sun—Miscellanea.] I. Title. II. Library of why?
QB521.5 .M87 2003
523.7—dc21

                              2001005469

Manufactured in the United States of America

Contents

What Is the Sun?

The Sun is the star in the center of our **solar system**. Without it days on Earth would be cold and dark. Plants would not grow. Humans would not live. Nothing would live. All living things need the Sun. The Sun is also a star in the Milky Way **galaxy**. Although it is the closest star to Earth, it is still 93 million miles (150 million km) away. The Sun measures 870,000 miles (1.4 million km) around. It is made of hot gases. These gases include hydrogen and helium.

◀ *The Sun lights our days and keeps us warm. It also helps living things to grow.*

How Was the Sun Formed?

Scientists believe that the Sun began four and a half billion years ago as a cloud of dust and gas. This cloud was a **solar nebula**. The solar nebula **rotated** around and around. It became a flat disk. Soon the cloud's **temperature** and **pressure** grew. Its center separated and formed the Sun. Other parts of the cloud became planets, asteroids, and comets. Asteroids are very small planets that orbit the Sun. Comets are small, rocky bodies with tails of dust, gas, and ice. Planets, asteroids, and comets all **orbit** the Sun. Earth's orbit around the Sun takes one year, or 365 days.

This drawing shows how planets may have been formed from a solar nebula. ▶

FORMATION OF PLANETS FROM A SOLAR NEBULA

How Does the Sun Get Around?

The Sun rotates, or spins, on an axis. As the Sun rotates, it orbits the Milky Way galaxy. While it orbits, it pulls along our solar system's nine planets, 63 moons, many asteroids, and comets.

It is the Sun's **gravity** that pulls and keeps these bodies orbiting in space. The Sun's gravity is 28 times greater than the Earth's gravity. The Sun has the strongest gravity in the solar system. This is because it has the most **mass**.

◄ *This illustration shows planets and a comet revolving around the Sun.*

Why Is the Sun So Hot?

The Sun is on fire! The Sun's top layer is called the **photosphere**. It is the layer that shines. Its boiling gases reach 10,500°F (5,815.5°C). There are cooler, darker parts on the Sun's surface, called sunspots. Scientists can see sunspots only with special telescopes.

The center of the Sun has temperatures as high as 27 million°F (15 million°C). The pressure in the center is so great that it changes the Sun's hydrogen into helium. This change causes solar energy, which creates the Sun's heat and light.

The core is the center of the Sun. The corona is the Sun's outer atmosphere. The chromosphere is an invisible layer. Solar flares are bursts of energy. Solar wind blows out from the corona. ▶

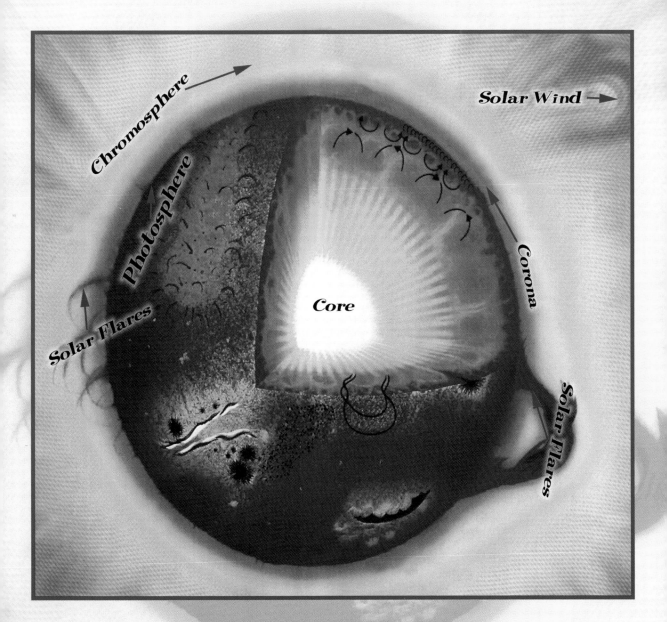

What Causes a Solar Eclipse?

Every two to four years, people somewhere on Earth can watch a solar eclipse. When the Moon's orbit passes in front of the Sun, the Moon blocks the Sun's light from Earth. This is called a partial solar eclipse. Only a part of the Sun's outer atmosphere, or corona, can be seen. During a total solar eclipse, only the Sun's corona can be seen. A total solar eclipse can last for as long as 7 ½ minutes. People should never look right at a solar eclipse. The brightness of the Sun can be blinding to human eyes.

◀ *This is a photograph of a total solar eclipse.*

What Is the Ozone Layer?

The ozone layer is a blanket of gas in Earth's atmosphere. The ozone layer is from 12 to 30 miles (19–48 km) thick. It guards people and animals on Earth from the Sun's harmful **ultraviolet rays**. These rays can cause sunburn. Over time they can even cause skin cancer, a harmful growth on the skin. Scientists have found holes in the ozone layer. These holes let through larger amounts of the Sun's ultraviolet rays. Scientists discovered that certain kinds of manufactured chemicals can cause these holes.

This special satellite view of Earth shows different levels of ozone gas. ▶

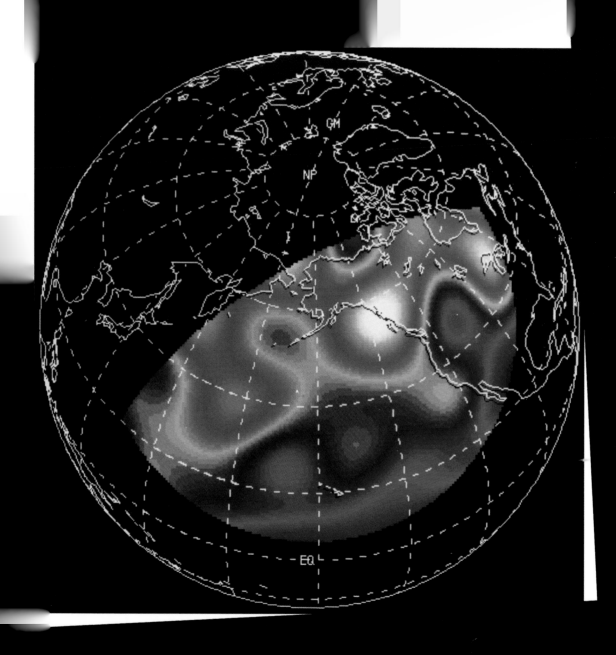

What Did People Once Believe About the Sun?

Early humans did not know what the Sun was. All they knew was that the Sun rose in the morning and set at night. They also knew it kept them warm and helped them to grow food. Some early people thought that the Sun was a special god. They gave it special offerings, or gifts, and thanked their god for shining! Early people also used the Sun to keep track of their days. When the Sun came up, they marked a new day. The Sun was Earth's first calendar.

◄ *Early Greeks used the Sun's shadow to tell time. They made sundials, or sun clocks, to chart the Sun's shadows.*

Will the Sun Last Forever?

The Sun is four and a half billion years old. Scientists believe that the Sun will last from five to seven billion years more. As the Sun uses up its **energy**, it becomes bigger and redder. It will turn into a red giant, an old star that has begun to die. The Sun's mass will grow bigger than the orbits of the planets and will swallow all of the planets in the solar system, one at a time. Using its last bit of energy, it becomes a white dwarf star, an old star that shrinks into a tiny ball. Soon the white dwarf will cool down, fade out, and die.

The Sun and all of the stars in the universe will eventually become red giants and white dwarfs. ▶

How Do Scientists Study the Sun?

Scientists study the Sun to learn about the stars and the solar system. They also study how the changes in the Sun affect life on Earth. To study the Sun, scientists use images from solar telescopes and spacecraft. Solar telescopes use mirrors to display pictures of the Sun on flat surfaces. These telescopes stand on mountains pointed toward the sky. The most powerful solar telescopes are called **spectroscopes**. They record the beams of light from the Sun. Scientists can tell things about the Sun from studying these beams.

◀ *Machines on spacecraft study the Sun. They have special telescopes, cameras, and computers to collect information.*

How Can People Keep Their Skin Safe from the Sun?

It is important to avoid the Sun's ultraviolet rays. We can do a lot of things to protect our skin from the Sun's ultraviolet rays. Stay out of the bright sunlight from 12:00 P.M. to 2:00 P.M. These are the hours when the Sun shines the strongest. Wear sunscreen on both sunny and cloudy days! It is important to reapply sunscreen after swimming or exercising. Wear a hat with a brim that is at least 4 inches (10 cm long). Wear sunglasses. The Sun can be harmful if we're not careful, but it's our friend, too. We could not live without it.

Glossary

energy (EH-nur-jee) The power to work or to act.

galaxy (GA-lik-see) A large group of stars and the planets that circle them.

gravity (GRA-vih-tee) The natural force that causes objects to move or tend to move toward each other.

mass (MAS) The amount of matter in something.

orbit (OR-bit) To travel a circular path around the Sun.

photosphere (FOH-toh-sfeer) The fourth layer of the Sun.

pressure (PREH-shur) A force that pushes on something.

rotated (ROH-tayt-ed) Moved in a circle.

solar nebula (SOH-ler NE-byuh-luh) A cloud of dust and gas that was the birthplace of the Sun.

solar system (SOH-ler SIS-tem) A group of planets that circles a star.

spectroscopes (SPEK-troh-skohps) Powerful telescopes used to study the Sun.

temperature (TEM-pruh-chur) How hot or cold something is.

ultraviolet rays (ul-truh-VY-let RAYZ) Rays of heat and light from the Sun that can be dangerous to humans.

Index

Web Sites

Due to the changing nature of Internet links, PowerKids Press has developed an online list of Web sites related to the subject of this book. This site is updated regularly. Please use this link to access the list:
www.powerkidslinks.com/low/sunhot/